BACK FROM THE BRINK

REPTILES

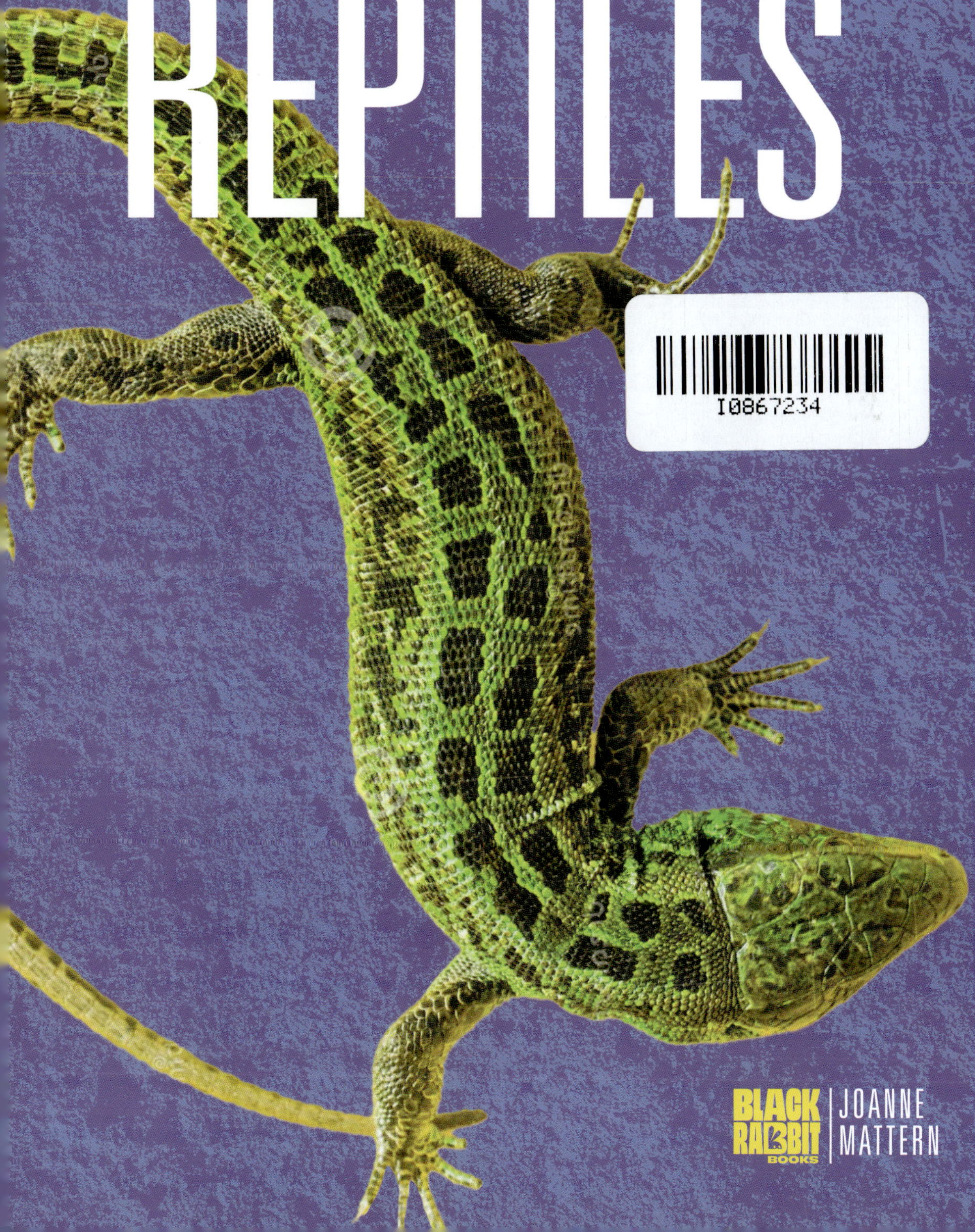

I0867234

BLACK RABBIT BOOKS

JOANNE MATTERN

TABLE OF CONTENTS

1 Blue Iguana

The blue iguana once lived all over the Cayman Islands. Then people arrived. They brought their cats and dogs. These pets killed the iguanas. Other blue iguanas were hit by cars. Over time, less than 25 were left.

People wanted to save this **reptile**. The government set aside land for them. Special boxes gave them a safe place to live. The boxes **protected** them from other animals. Today, more than 1,000 blue iguanas live in peace.

Think About It
How do people and their pets hurt wild animals?

2

Chinese Alligator

Chinese alligators lived in swampy areas in China. Then people drained the swamps. They used these areas for farming. By the 1970s, there were just a few alligators left.

The alligators were caught and sent to zoos. They had babies. Then they were put back into the wild. Today, the Chinese alligator is still rare. But it is slowly coming back.

Did You Know?

The Chinese alligator also lives in tunnels underground.

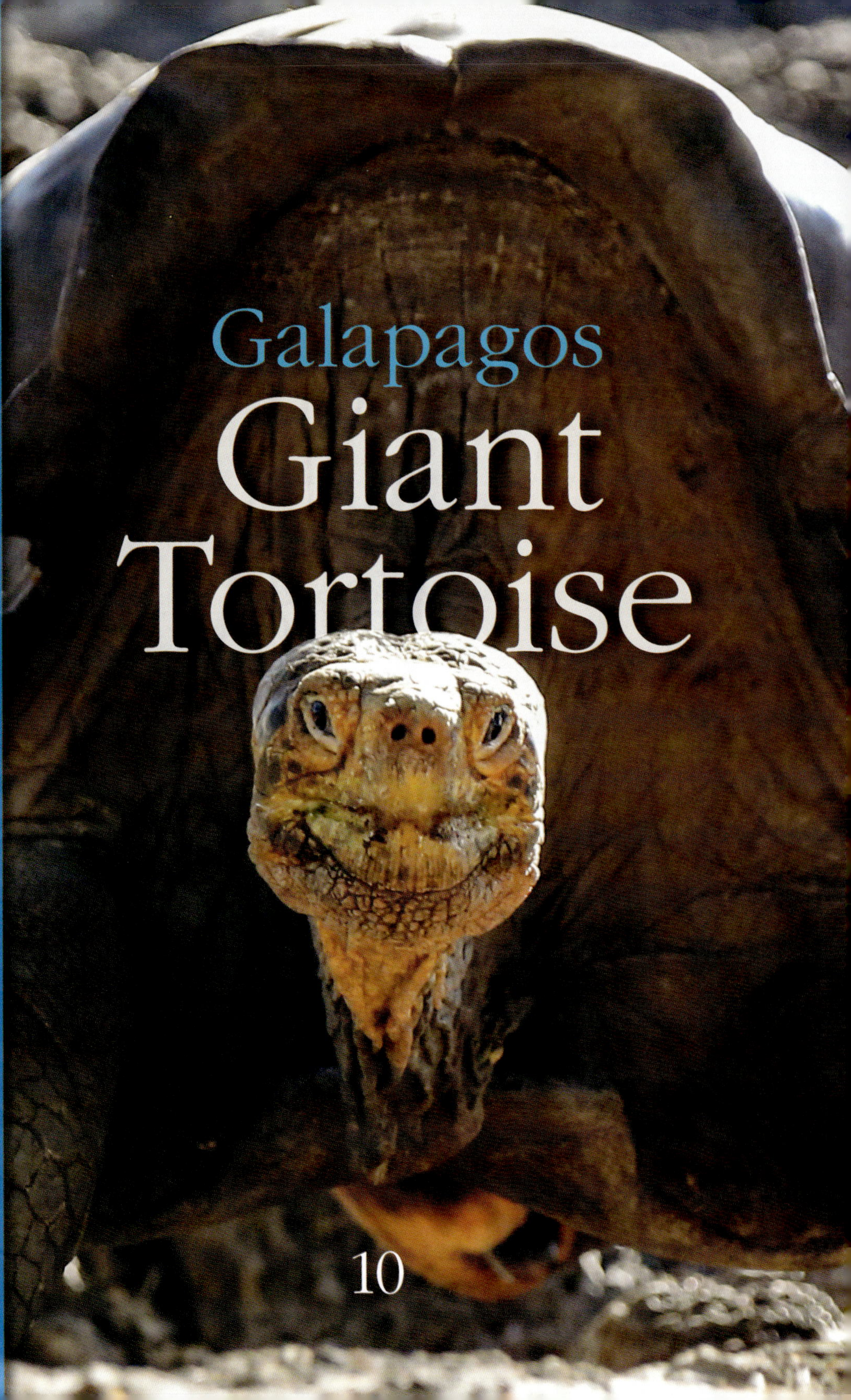

3

Galapagos Giant Tortoise

These tortoises were once common on the Galapagos Islands. But then people brought rats and dogs. These animals ate the tortoises' eggs. It seemed that the reptile would soon die out.

Scientists began gathering tortoise eggs. They raised the babies in safe spaces. The government set aside land for the tortoises to live. Today, these animals once again live in their island homes.

Did You Know?
This giant tortoise can live more than 100 years.

4

Kemp's Ridley Sea Turtle

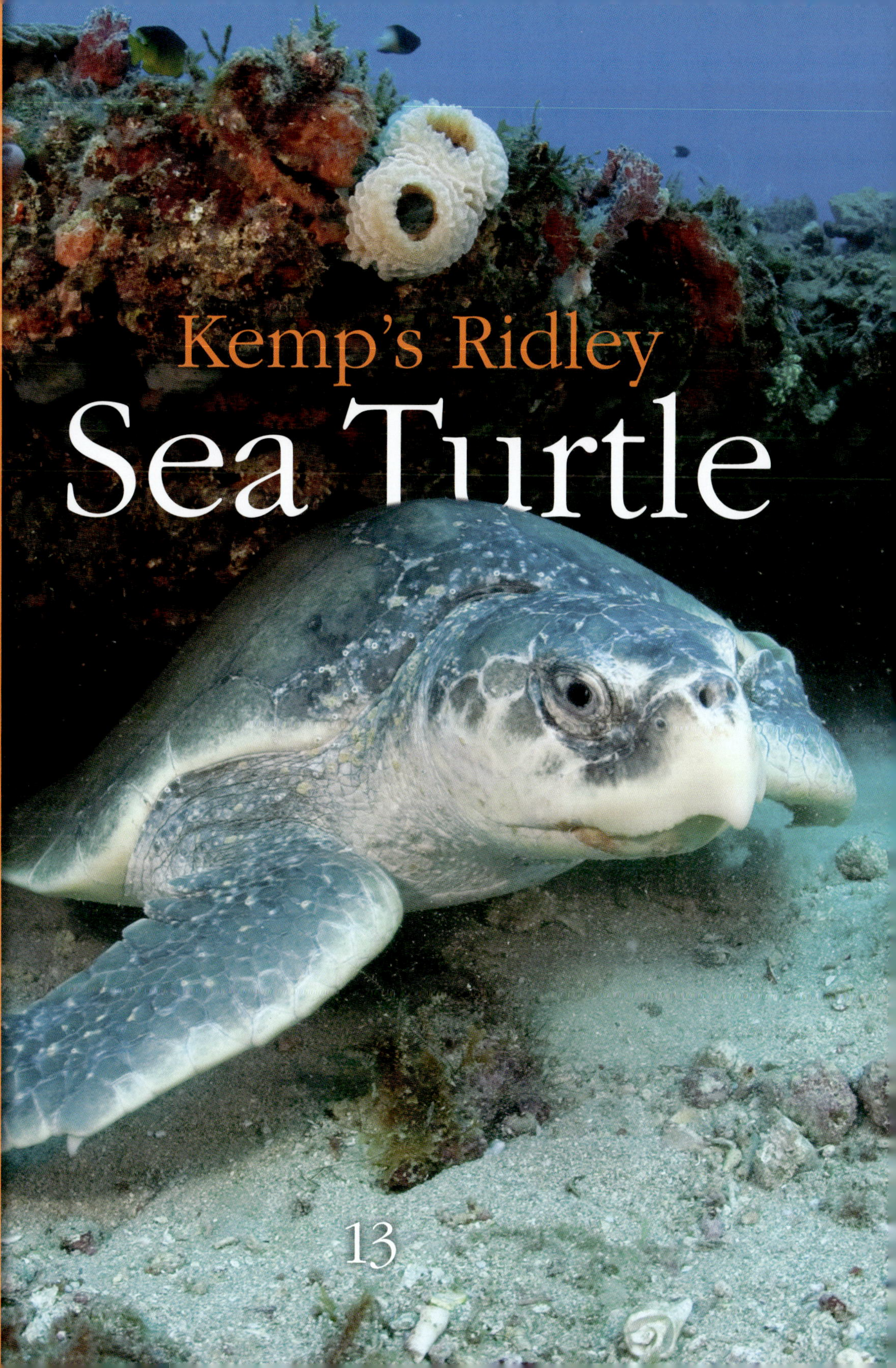

The Kemp's ridley sea turtle once lived all over the Gulf of Mexico. The turtles laid their eggs on beaches. But then people destroyed these beaches. They drove over the sand. They built houses on the beaches too. Many eggs and baby turtles were destroyed.

The United States and Mexico worked together to save the sea turtles. They stopped people from going to some beaches. Zoos gathered eggs and raised the babies. Today, thousands of sea turtles swim in the Gulf waters.

Did You Know?
This turtle is named after the fisherman who discovered them in 1906.

5

Lake Erie Water Snake

These snakes live on the islands of Lake Erie. There were once thousands. Then people arrived. They killed most of the snakes.

In 1986, the water snake was listed as **endangered**. Scientists protected places where the snake lived. They taught people that the snakes were an important part of the **ecosystem**. People listened. They left the snakes alone. In 2011, the snake was taken off the list of endangered animals.

Think About It

How can teaching people help save animals that are at risk?

Sand Lizard

6

Sand lizards live across Europe and Russia. They need large, sandy spots to lay their eggs. When people destroyed these places, the lizards died out.

The government stepped in to help. They protected the sand **dunes**. The lizards were raised in zoos. Then they were returned to the wild. Today, the lizards are back in the sandy homes they need.

Think About It
What role does the government play in helping animals at risk?

MORE TO EXPLORE

WHERE IN THE WORLD?

BLUE IGUANA
Cayman Islands

GALAPAGOS GIANT TORTOISE
Galapagos Islands

LAKE ERIE WATER SNAKE
Islands in Lake Erie

KEMP'S RIDLEY SEA TURTLE
Gulf of Mexico and Atlantic Ocean

SAND LIZARD
Europe and central Asia

CHINESE ALLIGATOR
China

MORE TO EXPLORE

FANTASTIC FACTS

The blue iguana is one of the longest-living types of lizards. They can live up to 40 years!

Chinese alligators are great swimmers and fast runners.

The Galapagos Islands were named after the giant tortoise.

Unlike other snakes, Lake Erie water snakes are not afraid of people.

Kemp's ridley sea turtles are the smallest sea turtles.

Male sand lizards turn bright green in the spring.

MORE TO EXPLORE

COOL COMPARISONS

These animals are making a comeback. How many live in the wild today?

Blue Iguana
about 1,000

Chinese Alligator
about 150

Galapagos Giant Tortoise
about 15,000

Kemp's Ridley Sea Turtle
7,000 to 9,000

Lake Erie Water Snake
about 10,000

Sand Lizard
about 9,000

MORE TO EXPLORE

RESOURCES

Glossary

dune (DOON) A hill or mound of sand piled up by the wind.

ecosystem (EE-co-sys-tuhm) A community of living things in one place.

endangered (en-DAYN-jurd) Close to no longer existing.

protect (pruh-TEKT) To defend or keep safe; shield from danger or harm.

reptile (REP-tahyl) A cold-blooded animal that breathes air and has a backbone; most reptiles lay eggs and have scaly skin.

swamp (SWOMP) Wet, spongy land that is partly covered with water.

Read More

Golkar, Golriz. *Endangered Reptiles Around the World*. North Mankato, MN: Pebble, 2025.

Koestler-Grack, Rachel A. *Giant Tortoises*. Minneapolis: Bellwether Media, 2023.

Index

TOP RANK is published by Black Rabbit Books, P.O. Box 227, Mankato, MN, 56002.

• COPYRIGHT © 2026 Black Rabbit Books. All rights reserved. No part of this book may be reproduced in any form without written permission from the publisher.

• Top Rank is an imprint of Black Rabbit Books • Edited by Ana Brauer • Series designed by Danny Nanos • Book designed by Jason Knudson • Photographs © Alamy Stock Photo/Alfred Schauhuber, 20, Michael Patrick O'Neill, 13, 14–15, 21, 23, Steve Paddon, 16, 21, 23; Dreamstime/Liuqf, 8, Pavel Trankov, cover, 1, Serban Enache, 5; Flickr/Nathan Rupert, 6–7, 21, 23; Shutterstock/Eva_Puhova, cover, 1, Ian Kennedy, 11, Jukka Palm, 17, Kurit afshen, 4, 21, 23, Lauren Suryanata, 9, Ryan M. Bolton, 2, 18, Todamo, 10, Viatore, 19, 21, 23; Wikimedia Commons/FelixReyesPhotography, 12, 21, 23 • Printed in India

Library of Congress Cataloging-in-Publication Data: Names: Mattern, Joanne, 1963- author. | Title: Reptiles: back from the brink / by Joanne Mattern. | Description: Mankato, MN: Top Rank, an imprint of Black Rabbit Books, [2026] | Series: Back from the brink | Includes bibliographical references and index. | Audience: Ages 8–11 | Audience: Grades 2–3 | Identifiers: LCCN 2024043403 (print) | LCCN 2024043404 (ebook) | ISBN 9781644668009 (library binding) | ISBN 9781644668320 (paperback) | ISBN 9781644668641 (ebook) | Subjects: LCSH: Reptiles—Juvenile literature. | Endangered species—Juvenile literature. | Classification: LCC QL644.7 M38 2026 (print) | LCC QL644.7 (ebook) | DDC 597.9—dc23/eng/20241223 | LC record available at https://lccn.loc.gov/2024043403 | LC ebook record available at https://lccn.loc.gov/2024043404